Cellular Manufacturing

SHOPFLOOR SERIES

Cellular Manufacturing
One-Piece Flow for Workteams

CREATED BY

**The Productivity
Development Team**

Productivity Press
PORTLAND, OREGON

Additional copies of this book and a learning package for leading a book study group are available from the publisher. Discounts are available for multiple copies through the Sales Department (800-394-6868). Address all other inquiries to:

Productivity Press
P.O. Box 13390
Portland OR 97213-0390
United States of America
Telephone: 503-235-0600
Fax: 503-235-0909
E-mail: service@productivityinc.com

Cover by Shad Lindo
Cover illustration by Gary Ragaglia
Page design and composition by William H. Brunson, Typography Services
Printed and bound by Malloy Lithographing, Inc. in the United States of America

Library of Congress Cataloging-in-Publication Data

Cellular manufacturing : one-piece flow for workteams / created by the Productivity
    Development Team.
        p. cm.
      ISBN 1-56327-213-X
      1. Assembly-line methods—Automation.   2. Manufacturing cells.
3. Production planning.   I. Productivity Development Team (Productivity Press)
    TS178.4.C45 1999
    658.5'1—dc21   98–50439
                                                                              CIP

04 03 02   10 9 8 7 6 5 4

Contents

Chapter 4. Teamwork Tools for Cellular Manufacturing   49

Chapter 5. Reflections and Conclusions   63

About the Productivity Development Team   69

Publisher's Message

Cellular manufacturing is an aspect of lean production that can dramatically boost your company's competitiveness by eliminating waste from the manufacturing process. By arranging people and equipment into efficient, process-based cells, cellular manufacturing creates a smooth flow that shortens the lead time for delivery to customers. It also supports low-inventory production of a variety of products to meet customer needs.

The change from traditional large-lot production to lean manufacturing requires a new understanding about what activities add value for the customer. Although management determines the manufacturing approach, frontline employees play a key role in making it work. This book is intended to share basic knowledge that will help you participate effectively as your company implements cellular manufacturing and other lean production approaches.

This book shares some topics with another Shopfloor Series book, *Just-in-Time for Operators*. That book also discusses lean production methods, but it is more about the "mechanics" of low-inventory production, such as takt time, load leveling, pull systems, and kanban. *Cellular Manufacturing* focuses on the team-based experience of changing to cellular manufacturing. Chapter 1 lays a foundation with basic definitions related to cellular manufacturing and the reasons it is so important for U.S. companies and their employees. Chapter 2 describes how the cellular arrangement changes the way people work—for example, learning new skills to handle several workstations in a sequence.

Chapter 3 offers a step-by-step overview of the process of converting a work area into a manufacturing cell. You will learn about observing the process to understand how it currently works and to identify wastes such as excess transport and waiting. You'll also be exposed to key tools of cell design, including standard work forms, PQ analysis, and documentation of the new procedure. Finally, you'll see how cell conversion is the beginning point for continuous improvement to eliminate bottlenecks, breakdowns, and other problems that block a smooth flow through the process.

Cellular manufacturing requires a team-based, cooperative environment. Chapter 4 describes key teamwork tools for keeping cells running smoothly. These include the 5S system for workplace organization and standardization, visual management tools, quick changeover and mistake-proofing approaches, and total productive maintenance—important techniques that make cell operation possible. Chapter 5 helps you review your learning and suggests additional resources for exploring key topics.

It is important to remember as you read that this material is a general orientation to a complex topic. Implementation and mastery of cellular manufacturing requires a deeper understanding of the production mechanism. The implementation process is best supported by experienced consultants and trainers who can tailor it to your company's specific situation and help address the issues that may be raised by this change.

This book incorporates a number of features that will help you get the most from it. Each chapter begins with an overview of the contents. The book uses many illustrations to share information and examples in a visual way. Icon symbols in the margin flag key points to remember in each section. And questions built into the text provide a framework for applying what you've learned to your own situation.

One of the most effective ways to use this book is to read and discuss it with other employees in group learning sessions. We have deliberately planned the book so that it can be used this way, with chunks of information that can be covered in a series of short sessions. Each chapter includes reflection questions to stimulate group discussion. A Learning Package is also available, which includes a leader's guide, overhead transparencies for major points, and color slides showing examples of cellular manufacturing techniques and experiences in different companies.

The cellular manufacturing approach works in companies all over the world. Today, the basic principles of cell design have been used to eliminate waste in all types of manufacturing, assembly, and even service industries. We hope this book and Learning Package will tell you what you need to know to make your participation in a cellular manufacturing implementation active and personally rewarding.

Acknowledgments

The development of *Cellular Manufacturing* has been a team effort. Productivity Consulting Group members Frank Hammitt and Richard Niedermeier served as content advisors. Cover composition was by Shad Lindo and cover illustration by Gary Ragaglia of The Vision Group. Lorraine Millard managed the prepress production and manufacturing, with editorial assistance from Pauline Sullivan. Page composition and additional design was done by William H. Brunson Typography Services. Graphic illustrations were created by Lee Smith and cartoon illustrations were created by Hannah Bonner.

Finally, the Productivity staff wishes to acknowledge the good work of the many people who are in the process of implementing cellular manufacturing in their own organizations. We welcome your feedback about this book, as well as input about how we can continue to serve you in your implementation efforts.

Steven Ott
President

Karen Jones
Productivity Development Team

Getting Started

The Purpose of This Book

Key Point

Cellular Manufacturing *was written to give you the information you need to participate in implementing this important lean manufacturing approach in your workplace.* You are a valued member of your company's team; your knowledge, support, and participation are essential to the success of any major effort in your organization.

The paragraph you have just read explains the author's purpose in writing this book. It also explains why your company may wish you to read this book. But why are you reading this book? This question is even more important. What you get out of this book largely depends on your purpose in reading it.

You may be reading this book because your team leader or manager asked you to do so. Or you may be reading it because you think it will provide information that will help you in your work. By the time you finish Chapter 1, you will have a better idea of how the information in this book can help you and your company eliminate waste and serve your customers more effectively.

What This Book Is Based On

BACKGROUND
INFO

This book is about an approach to arranging the workplace for processing items with minimal waste and delay. The methods and goals discussed here are closely related to the lean manufacturing system developed at Toyota Motor Company. Since 1979, Productivity, Inc. has brought information about these approaches to the United States through publications, events, training, and consulting. Today, top companies around the world are applying lean manufacturing principles to sustain their competitive edge.

Cellular Manufacturing draws on a wide variety of Productivity's resources. Its aim is to present the main concepts and techniques of cellular manufacturing and one-piece flow in a simple, illustrated format that is easy to read and understand.

Figure I-1. Two Ways to Use This Book

Two Ways to Use This Book

BACKGROUND
INFO

There are at least two ways to use this book:

1. As the reading material for a learning group or study group process within your company.

2. For learning on your own.

Productivity Press offers a Learning Package that uses *Cellular Manufacturing* as the foundation reading material for a learning group. Your company may decide instead to design its own learning group process based on *Cellular Manufacturing*. Or, you may read this book for individual learning without formal group discussion.

How to Get the Most Out of Your Reading

Becoming Familiar with This Book as a Whole

There are a few steps you can follow to make it easier to absorb the information in this book. Take as much time as you need to become familiar with the material. First, get a "big picture" view of the book by doing the following:

How-to Steps

1. Scan the Table of Contents to see how *Cellular Manufacturing* is arranged.

2. Read the rest of this section for an overview of the book's contents.

3. Flip through the book to get a feel for its style, flow, and design. Notice how the chapters are structured and glance at the pictures.

Becoming Familiar with Each Chapter

After you have a sense of the structure of *Cellular Manufacturing*, prepare yourself to study one chapter at a time. For each chapter, we suggest you follow these steps to get the most out of your reading:

How-to Steps

1. Read the "Chapter Overview" on the first page to see where the chapter is going.

2. Flip through the chapter, looking at the way it is laid out. Notice the bold headings and the key points flagged in the margins.

3. Now read the chapter. How long this takes depends on what you already know about the content, and what you are trying to get out of your reading. Enhance your reading by doing the following:

   • Use the margin assists to help you follow the flow of information.

   • If the book is your own, use a highlighter to mark key information and answers to your questions about the material. If the book is not your own, take notes on a separate piece of paper.

   • Answer the "Take Five" questions in the text. These will help you absorb the information by reflecting on how you might apply it at work.

4. Read the "Chapter Summary" to confirm what you have learned. If you don't remember something in the summary, find that section in the chapter and review it.

5. Finally, read the "Reflections" questions at the end of the chapter. Think about these questions and write down your answers.

Figure I-2. Giving Your Brain a Framework for Learning

How a Reading Strategy Works

When reading a book, many people think they should start with the first word and read straight through until the end. This is not usually the best way to learn from a book. The steps described on pages xiv and xv are a strategy for making your reading easier, more fun, and more effective.

Key Point

Reading strategy is based on two simple points about the way people learn. The first point is this: *It's difficult for your brain to absorb new information if it does not have a structure to place it in.* As an analogy, imagine trying to build a house without first putting up a framework.

Like building a frame for a house, you can give your brain a framework for the new information in the book by getting an overview of the contents and then flipping through the materials. Within each chapter, you repeat this process on a smaller scale by reading the overview, key points, and headings before reading the text.

Key Point

The second point about learning is this: *It is a lot easier to learn if you take in the information one layer at a time, instead of trying to absorb it all at once.* It's like finishing the walls of a house: First you lay down a coat of primer. When it's dry, you apply a coat of paint, and later a final finish coat.

Using the Margin Assists

As you've noticed by now, this book uses small images called
margin assists to help you follow the information in each chapter.
There are six types of margin assists:

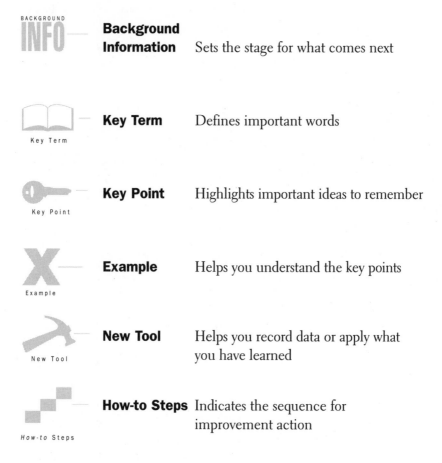

Background Information — Sets the stage for what comes next

Key Term — Defines important words

Key Point — Highlights important ideas to remember

Example — Helps you understand the key points

New Tool — Helps you record data or apply what you have learned

How-to Steps — Indicates the sequence for improvement action

Overview of the Contents

Getting Started (pages xiii – xviii)

This is the section you're reading now. It explains the purpose of
Cellular Manufacturing and how it was written. Then it shares
tips for getting the most out of your reading. Finally, it presents
this overview of each chapter.

Chapter 1. An Introduction to Cellular Manufacturing (pages 1–11)

Chapter 1 introduces and defines cellular manufacturing and the kinds of waste it helps eliminate. It also explains how the cellular manufacturing approach benefits companies and their employees, and defines processes and operations as a framework for the rest of the book.

Chapter 2. Working in a Manufacturing Cell (pages 13–21)

Chapter 2 describes important features you may experience when you work in a cell, including a process-based equipment layout, cross-training for multimachine operation, smaller and simpler machines, and autonomation (*jidoka*).

Chapter 3. Basic Elements of Cell Design (pages 23–47)

Chapter 3 highlights key steps in the three basic phases of converting a process into a manufacturing cell: understanding current conditions, making the conversion, and continuous improvement. Examples of various information-gathering forms are included.

Chapter 4. Teamwork Tools for Cellular Manufacturing (pages 49–61)

Chapter 4 covers essential methods that support teamwork in the new workplace arrangement, including the 5S system, visual management, and autonomous maintenance activities. Activity boards and one-point lessons are described as information-sharing tools for the team.

Chapter 5. Reflections and Conclusions (pages 63–68)

Chapter 5 presents reflections on and conclusions to this book. It also describes opportunities for further learning about cellular manufacturing and related techniques.

Chapter 1

An Introduction to Cellular Manufacturing

Type	Examples
Defects	Scrap, rework, replacement production, inspection
Waiting	Stockouts, lot processing delays, equipment downtime, capacity bottlenecks
Processing	Unnecessary or incorrect processing
Overproduction	Manufacturing items for which there are no orders
Movement	Human motions that are unnecessary or straining
Inventory	Excess raw material, WIP, or finished goods
Transport	Carrying WIP long distances, inefficient transport
Unused employee creativity	Lost time, ideas, skills, improvements

Figure 1-1. Waste in Manufacturing

What Is Cellular Manufacturing?

Key Term

Key Point

Cellular manufacturing is a lean manufacturing approach that helps companies build a variety of products for their customers with as little waste as possible. *In cellular manufacturing, equipment and workstations are arranged in a sequence that supports a smooth flow of materials and components through the process, with minimal transport or delay.*

Key Term

Key Term

Cellular manufacturing is a major building block of lean manufacturing. *Lean manufacturing* is an approach for building the variety of products customers require, profitably. Lean manufacturing makes companies more profitable and competitive by reducing wastes that typically add cost and lead time to the manufacturing process. *Waste* in this sense means any element of the manufacturing process that adds cost without adding value to the product. Figure 1-1 lists eight types of waste addressed by a lean manufacturing system.

Key Term

Cellular manufacturing gets its name from the word *cell*. A *manufacturing cell* consists of the people and the machines or workstations required for performing the steps in a process or process segment, with the machines arranged in the processing sequence. For example, if the process for a particular product requires cutting, followed by drilling and finishing, the cell would include the equipment for performing those steps, arranged in that order.

Figure 1-2. Loading a Machine with One Piece of WIP

Arranging people and equipment into manufacturing cells helps companies achieve two important goals of lean manufacturing— one-piece flow and high-variety production.

One-Piece Flow

One-piece flow is the state that exists when products move through a manufacturing process one unit at a time, at a rate determined by the needs of the customer (see Figure 1-2).

The opposite of one-piece flow is *large-lot production*. Although many companies produce goods in large lots or batches, that approach to production builds delays into the process. No items can move on to the next process until all the items in the lot have been processed. The larger the lot, the longer the items sit and wait between processes.

Key Term

Key Term

Figure 1-3. Large-Lot Production Creates Waste

Large-lot production can lower a company's profitability in several ways (see Figure 1-3).

- It lengthens the lead time between the customer's order and delivery of the products.

- It requires labor, energy, and space to store and transport the products.

- It increases the chances of product damage or deterioration.

In contrast, one-piece flow production solves these problems.

Key Points

- It allows the company to deliver a flow of products to customers with less delay.

- It reduces the resources required for storage and transport.

- It lowers the risk of damage, deterioration, or obsolescence.

- It exposes other problems so they can be addressed.

Key Point

One-piece flow is an ideal state; in daily operation, it is not always possible or desirable to process items just one at a time. *The important thing is to promote a continuous flow of products, with the least amount of delay and waiting.* Cellular manufacturing helps by focusing on the material going through the process, not just on the equipment for each operation.

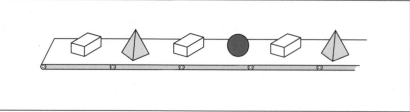

Figure 1-4. High-Variety Production

High-Variety Production

In the early days of industry, a company could produce one type of product and customers would buy it, even if it wasn't exactly what they would have liked. Today, however, customers expect variety and even customization, as well as specific quantities delivered at a specific time. If your company isn't flexible enough to serve their needs, they will go to your competitor.

Key Point

Cellular manufacturing offers companies the flexibility to give customers the variety they want (see Figure 1-4). It does this by grouping similar products into families that can be processed on the same equipment in the same sequence. It also encourages companies to shorten the time required for changeover between products. This eliminates a major reason for making products in large lots—that changeovers take too long to change the product type frequently.

TAKE FIVE

Take five minutes to think about these questions and to write down your answers:

- What is the size of a typical processing lot or batch in your work area?
- How many different types of products does your work area produce in a typical week? In a typical day?

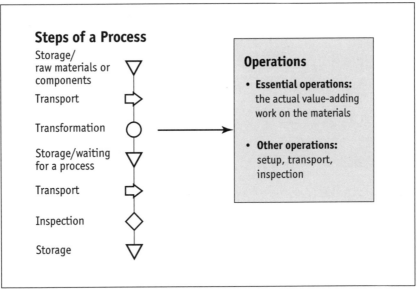

Steps of a Process

Storage/
raw materials or
components

Transport

Transformation

Storage/waiting
for a process

Transport

Inspection

Storage

Operations

• **Essential operations:**
the actual value-adding
work on the materials

• **Other operations:**
setup, transport,
inspection

Figure 1-5. Processes and Operations

Understanding Processes and Operations

Converting a factory to cellular manufacturing means eliminating waste from processes as well as from operations. It is important to understand how these two basic elements differ and where they intersect.

Processes

Key Term

A *process* is a continuous flow through which raw materials are converted to finished products in a series of operations. The focus is the path of the materials as they are transformed into something to sell.

Manufacturing processes have four basic types of steps or phases:

Key Points

- **Transformation:** assembly, disassembly, alteration of shape or quality
- **Inspection:** comparison with a standard
- **Transport:** change of location
- **Storage:** a waiting period when nothing else is happening

Materials and parts often go through several of these steps during a manufacturing process; the left side of Figure 1-5 shows a typical sequence of process steps. *Notice that only the transformation step adds value to the product.*

Key Point

Operations

Key Term

In contrast to a process, which focuses on flow, an operation focuses on action. An *operation* is any action performed by workers or machines on the raw materials, work-in-process, or finished products. The right side of Figure 1-5 gives examples of manufacturing operations.

Since operations involve actions, operational improvements focus on how specific actions are carried out. Operation improvement approaches include studying the motions required for a specific action, adjusting the height or angle of a work surface for easier use, and so on.

Key Point

To improve production for lean manufacturing, it is not enough to improve operations. Companies must also improve their processes. Improving a process involves streamlining the flow of materials to minimize obstacles and wastes such as

- Time spent in non-value-adding steps such as waiting or transport

- Downtime caused by changeover and adjustments

- Distance materials or WIP must travel between transformation steps

- The need for inspection, or for reworking materials

Key Point

The cellular manufacturing approach works on improving the process as well as specific operations.

TAKE FIVE

Take five minutes to think about these questions and to write down your answers:

- What kind of processes happen in your work area?
- What kind of operations do you perform in your daily work?

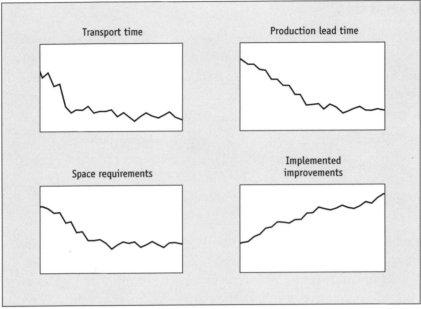

Figure 1-6. Cellular Manufacturing Improves Competitiveness

Benefits of Cellular Manufacturing

How It Helps Companies

Key Point

Promoting one-piece flow through cellular manufacturing can help make your company more competitive (see Figure 1-6). The cellular manufacturing approach

- Cuts costly transport and delay from the manufacturing process

- Shortens the production lead time, which serves customer needs and gives an earlier return on the company's investment in the product

- Saves space in the factory that can be used for other value-adding purposes

- Promotes continuous improvement by forcing solutions to problems that block low-inventory production

How It Helps You

Cellular manufacturing also helps you as a company employee. *By strengthening the company's competitiveness, it helps support job security.* In addition, it also makes daily production work go smoother by

Key Point

- Removing the clutter of excess WIP inventory

- Reducing transport and unnecessary handling

- Reducing the walking required to complete the process

- Addressing causes of defects and machine problems that cause delays

As your company implements cellular manufacturing, you may be asked to learn a process sequence you have not done before. This will raise your skill and flexibility, and it may change how you think about your role in the company. It is important to recognize that *learning about and participating in a JIT transformation ultimately will make you more employable.*

Key Point

TAKE FIVE

Take five minutes to think about these questions and to write down your answers:

- Based on what you have learned, can you see how cellular manufacturing might benefit your company? If so, how?
- Can you see how it might benefit you? If so, how?

In Conclusion

SUMMARY

Cellular manufacturing is a lean manufacturing approach that helps companies build a variety of products for their customers with as little waste as possible. In cellular manufacturing, equipment and workstations are arranged in a sequence that supports a smooth flow of materials and components through the process, with minimal transport or delay. Cellular manufacturing is a major building block of lean manufacturing.

A manufacturing cell consists of the people and the machines or workstations required for performing the steps in a process, with the machines arranged in the processing sequence. Arranging people and equipment into cells helps companies achieve one-piece flow and high-variety production.

One-piece flow is the state that exists when products move through a manufacturing process one unit at a time, at a rate determined by the needs of the customer. Operating so that products flow one piece at a time allows the company to deliver products to customers quicker, reduces storage and transport requirements, lowers the risk of damage, and exposes other problems to address.

Cellular manufacturing offers companies the flexibility to give customers the variety they want. It allows variety by grouping similar products into families that can be processed on the same equipment. It also encourages companies to shorten changeover time so product types can be changed more frequently.

Converting a factory to cellular manufacturing means eliminating waste from processes as well as from operations. A process is a continuous flow in which raw materials are converted to finished products in a series of operations. The focus of a process is the path of the materials as they are transformed into something to sell. An operation is any action performed by workers or machines on the raw materials, work-in-process, or finished products.

Cellular manufacturing can help make your company more competitive by cutting out costly transport and delay, shortening the production lead time, saving factory space that can be used for other value-adding purposes, and promoting continuous improvement by forcing the company to address problems that block low-inventory production.

Cellular manufacturing helps you as an employee by strengthening the company's competitiveness, which helps support job security. It also makes daily production work go smoother by removing the clutter of excess WIP inventory, reducing transport and handling, reducing the walking required, and addressing causes of defects and machine problems.

REFLECTIONS

Now that you have completed this chapter, take five minutes to think about these questions and to write down your answers:

- What did you learn from reading this chapter that stands out as particularly useful or interesting?

- Do you have any questions about the topics presented in this chapter? If so, what are they?

- What additional information do you need to fully understand the ideas presented in this chapter?

Chapter 2

Working in a Manufacturing Cell

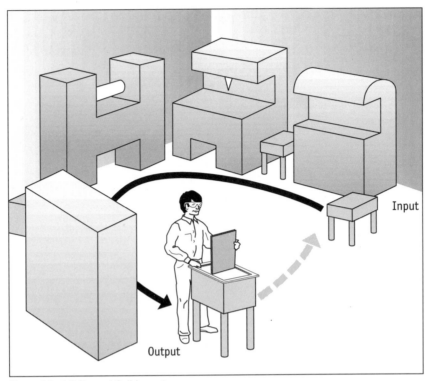

Figure 2-1. A U-Shaped Cell Layout

Changing from an operation-based layout to manufacturing cells will change how people do their work in the factory. This chapter describes some of the differences you may experience when working in a manufacturing cell.

Operating in a U-Shaped Cell

Key Point

In a manufacturing cell, the equipment and workstations are arranged close together in the sequence of the processing steps. This arrangement reduces unnecessary walking and transport to promote a smooth flow.

Furthermore, the equipment in a cell usually is laid out in a curved shape, so that the operator's path is like a U or a C (see Figure 2-1).

Key Point

These shapes bring the end point of the process close to the beginning point, which minimizes the distance the operator has to travel to begin the next cycle

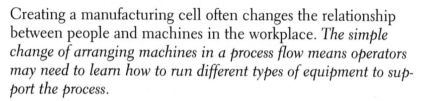

TAKE FIVE

Take five minutes to think about this question and to write down your answer:

• How much walking do you need to do in the course of processing the products?

Becoming Multiskilled, Multimachine Operators

Key Point

Creating a manufacturing cell often changes the relationship between people and machines in the workplace. *The simple change of arranging machines in a process flow means operators may need to learn how to run different types of equipment to support the process.*

In an operation-based layout, all the grinding machines, for example, would be located together. However, when the machines are rearranged into a cell according to the process sequence, each grinder may become part of a different cell. In that situation, having one grinder operator for each cell would not be economical. What's more, cells often use equipment that runs on automatic cycles, so most of the operator's time would be spent watching the equipment run. This is a huge waste of people's intelligence and skill.

Key Point

These wastes are avoided by teaching people to operate several different machines in the process. *With simple automation, an operator can manage the flow of work through a series of machines in the process.* For example, the operator can be setting up a workpiece on the equipment for step 2 while the step 1 machine is processing another workpiece.

A cell may be run by one person, or by several people working together, depending on the size of the cell, machine cycle times, and the production volume. This flexibility to change how people work together in a cell comes from cross-training.

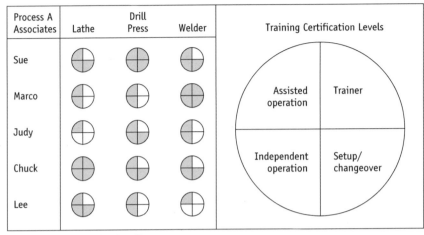

Figure 2-2. A Cross-Training Chart

Cross-Training for Maximum Flexibility

Key Point

Cross-training enables employees to perform different functions within a process and lets teams take full responsibility for their processes. When an operator is trained on several machines, he or she is qualified to respond to changes in production needs by stepping into other positions as needed. This versatility makes employees more valuable to their teams and to their companies.

Cross-training is a source of employee pride in many workplaces. Visual display charts are often used to recognize people's skill attainment in a public way (see Figure 2-2).

Moving with the Work

Key Point

To run several machines in sequence, an operator needs to work standing up rather than sitting down. In one-piece flow manufacturing, the work must move smoothly through the process. To assist this flow, people need to stand and walk. Working while standing also enables people to respond more quickly if machine problems occur.

TAKE FIVE

Take five minutes to think about these questions and to write down your answers:

- Do operators at your company run different kinds of machines? If not, what would need to happen to make this possible?
- Do operators at your company run more than one machine? If not, what would need to happen to make this possible?

Using Small, Flexible Machines

Key Point

A cellular manufacturing process may use equipment different from that used in large-lot production. *Cellular manufacturing works best with machines that are smaller and often slower than large-lot equipment.*

Smaller machines can be used for cellular manufacturing because the goal is to process one or a few items at a time, instead of large batches. Smaller machines save space. Placing them close together reduces the walking distance and leaves no space for excess WIP to accumulate.

Slower machines are appropriate for cellular manufacturing because the objective is not to produce large lots of WIP quickly. Instead, machines produce one piece at a time at a speed determined by customer requirements.

Key Point

Machines for cellular manufacturing also need to be flexible. To maximize their usefulness, they must be easy to set up quickly so they can be used to make a greater variety of products during a single shift. (See Chapter 3 for more about improving changeover time.)

Flexible may also mean movable. Mounting smaller machines on wheels makes it possible to move them to other locations when a process sequence changes, or to experiment with new production layouts.

Another benefit of using smaller machines for cellular manufacturing is that they generally are less expensive to purchase and easier to operate and maintain.

TAKE FIVE

Take five minutes to think about these questions and to write down your answers:

- Do you think the equipment in your area is better suited to large-lot production or to cellular manufacturing for one-piece flow? Why?

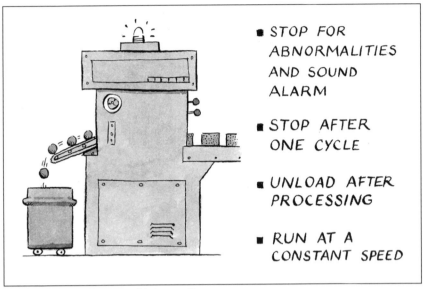

- STOP FOR ABNORMALITIES AND SOUND ALARM

- STOP AFTER ONE CYCLE

- UNLOAD AFTER PROCESSING

- RUN AT A CONSTANT SPEED

Figure 2-3. Typical Features of Autonomation

Using Autonomation (*Jidoka*) to Eliminate Machine Watching

Key Term

Another characteristic of the equipment used in just-in-time man-ufacturing is autonomation (also called *jidoka*). *Autonomation* is an approach to automation that gives equipment "intelligence" so people don't have to monitor automatic operation.

Key Point

"Autonomated" machines are semi-automatic machines that autonomously (independently) support one-piece flow processing. *They stop and signal when a cycle is complete or when problems occur* (see Figure 2-3). Although such machines are often loaded by operators, they are often set up to unload automatically after processing so they do not have to be tended.

Many companies invest in automated equipment so people don't have to perform difficult, dangerous, or repetitive work. At a lot of factories, however, people still watch the automated equipment "just in case" something goes wrong. Autonomation frees people from this non-value-adding role by modifying machines so they can run with little supervision.

The function of stopping for problems is also a key element of the mistake-proofing approach called poka-yoke. Poka-yoke systems are described further in Chapter 3.

The technology required for autonomation is often very simple. It is usually not expensive to modify existing machines to perform this way.

When people don't have to watch a machine to spot problems or to catch the output, they have more time for value-adding work or improvements, such as operating other machines or planning and implementing new ideas for improving the work flow.

TAKE FIVE

Take five minutes to think about these questions and to write down your answers:

- Do operators in your work area monitor automated equipment? If so, what are some of the things people must watch for?

In Conclusion

SUMMARY

In a manufacturing cell, the equipment and workstations are arranged close together in the sequence of the processing steps. This reduces unnecessary walking and transport to promote a smooth flow. The equipment usually is laid out in a curved shape like a U or a C. These shapes bring the end point of the process close to the beginning point, shortening the operator's travel to begin the next cycle.

Creating a manufacturing cell changes the relationship between people and machines in the workplace. Arranging machines in a process flow means operators may need to learn how to run different types of equipment to support the process.

An operation-based layout would group similar machines, such as grinders, in one part of the plant. However, when the machines are rearranged into a process-based cell, each grinder may become part of a different cell, and having one grinder operator for each cell would not be economical. Since cells often run equipment on automatic cycles, the operator's time, intelligence, and skill would be wasted in watching.

These wastes are avoided by teaching people to operate several different machines in the process. Simple automation allows an operator to manage the flow of work through a series of machines.

A cell may be run by one person, or by several people working together, depending on the situation. The flexibility to change how people work together in a cell comes from cross-training. Cross-training enables employees to perform different functions within a process and lets teams take full responsibility for their processes. It is a source of employee pride in many workplaces.

To run several machines in sequence, an operator needs to work standing up rather than sitting down. Working while standing also enables people to respond more quickly if machine problems occur.

Cellular manufacturing works best with machines that are smaller and often slower than large-lot equipment. Smaller machines save space. Placing them close together reduces walking and leaves no space for excess WIP to accumulate. Slower machines can be used because the aim is to produce one piece at a time at a speed determined by customer requirements. Machines for cellular manufacturing also need to be flexible — easy to set up quickly so they can be used for a greater variety of products.

Autonomation (*jidoka*) is an approach to automation that gives equipment "intelligence" so people don't have to monitor automatic operation. "Autonomated" machines are semi-automatic machines that autonomously (independently) support one-piece flow processing. They stop and signal when a cycle is complete or when problems occur. They are often set up to unload automatically after processing.

REFLECTIONS

Now that you have completed this chapter, take five minutes to think about these questions and to write down your answers:

- What did you learn from reading this chapter that stands out as particularly useful or interesting?

- Do you have any questions about the topics presented in this chapter? If so, what are they?

- What additional information do you need to fully understand the ideas presented in this chapter?

Chapter 3

Basic Elements of Cell Design

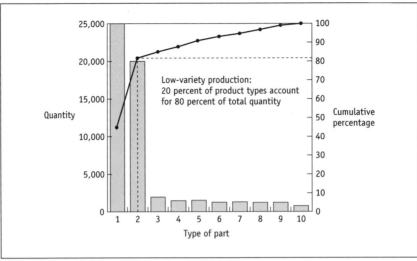

Figure 3-1. A PQ Analysis Chart Showing a 20:80 PQ Ratio (Low Variety)

This chapter* describes the three basic phases involved in converting a work area into a manufacturing cell:

1. Understanding the current conditions

2. Converting to a process-based layout

How-to Steps

3. Continuously improving the process

Phase 1: Understanding the Current Conditions

Key Point

The first phase in cell conversion, understanding the current conditions, helps the conversion team determine what process to convert. It also provides a baseline against which to measure improvement.

Collect Product and Production Data

Product Mix

New Tool

Deciding what process to convert to a cell is simpler if the company makes a few product types in relatively high volume. A high volume process tends to pull the maximum benefit from the improvement, while low variety avoids issues such as changeover. PQ (*product type/quantity*) analysis is used to display the product mix as a Pareto chart (see Figures 3-1 and 3-2).

---

*This chapter is longer than the other chapters. You may want to read it in sections; for example, you might read the text about one of the phases, then pause for reflection and group discussion.

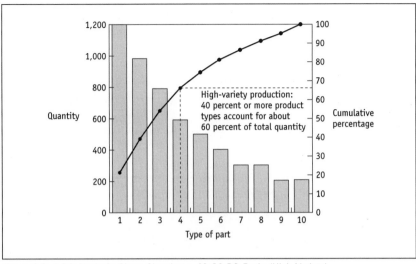

Figure 3-2. A PQ Analysis Chart Showing a 40:60 PQ Ratio (High Variety)

Key Point

PQ analysis makes lines on the chart to test the situation against several situations called PQ ratios. A high volume of a few product types appears as a 20:80 PQ ratio—where 20 percent of product types (horizontal axis) account for 80 percent of quantity (vertical axis) (see Figure 3-1). A 40:60 PQ ratio, on the other hand, indicates high variety of product types with relatively low volume of each type (see Figure 3-2). The team looks for the ratio where the second number is closest to the actual quantity percentage.

Production Resources

In addition to reviewing the product mix, the team gathers baseline information about production resources, such as

- shifts per day
- hours per shift; break time
- work days per month
- employee to operation ratio
- monthly product volume requirements from customers
- approach for assigning work
- finished goods inventory turns per month

TAKE FIVE

Take five minutes to think about this question and to write down your answer:

- What do you think your company's PQ analysis chart would look like?

Part code	Processes					
	1 Rough cut	2 Cut	3 Mill	4 Drill	4A Outside diameter	5 Gauge
A	⊙	⊙	⊙	⊙	⊙	⊙
D	⊙	⊙	⊙	⊙	⊙	⊙
C		⊙	⊙	⊙		⊙
I		⊙	⊙	⊙		⊙
G	⊙	⊙	⊙	⊙	⊙	
H	⊙	⊙	⊙	⊙	⊙	
J	⊙	⊙	⊙	⊙	⊙	
B	⊙	⊙		⊙	⊙	
E		⊙		⊙		⊙
F	⊙	⊙	⊙		⊙	

Figure 3-3. Process Route Analysis

Document Current Layout and Flow

Key Term

The next step in developing a baseline is understanding what operations make up the process and how that process is currently performed. This involves two activities: *process route analysis* (also called product route analysis) and *process mapping*.

Process Route Analysis

New Tool

A *process route analysis table* like the one in Figure 3-3 helps the team identify processing similarities between different products. This enables them to identify groups of products that could be made in a cell, using the same sequence of machines. These groups are called *product families*. If the company makes low volumes of many product types rather than high volumes of a few types, *process route analysis is especially important for helping the team choose a process to start with.*

Key Term

Key Point

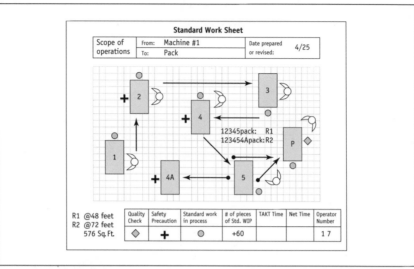

Figure 3-4. Standard Work Sheet

Process Mapping

Key Term

New Tool

Next, the team draws a *process map*, which shows the current equipment layout and the path the product takes through the process. This map is often drawn on a *Standard Work Sheet* (see Figure 3-4). The map also shows worker positions, WIP storage points, quality checkpoints, and safety precautions. In addition, the team records

- The distance the product must travel during processing
- The quantity of work-in-process in the process at a given time
- The number of people currently required to run the process

The team invites discussion with employees on the shop floor to learn about any special conditions that have not been documented. Through these steps it develops a thorough understanding of the basic elements of the current production process.

TAKE FIVE

Take five minutes to think about these questions and to write down your answers:

- Does your work area use the same machines to make different types of products?
- Would a map of your current equipment layout show a short, efficient path or one that zigzags around?

Process		TIME OBSERVATION SHEET															DATE	4/25	OBV.	FH
Machine #2																	Time	10a.m.	OBV.	RN
#	Component Task	1	2	3	4	5	6	7	8	9	10	11	12	13	14	15	TASK TIME	NOTES		
1	off timer	1/1																		
2	Rem/aside guard	3/2																		
3	Unload/aside part	6/3																		
4	Get/load part	9/3																		
5	Get/replace guard	11/2																		
6	Set timer/start	13/2																		
7	Assem/aside part	20/7																		
8	Wait (mach. cycle)	28/8																		
	TIME FOR ONE CYCLE	28																		

Figure 3-5. Time Observation Sheet

Time the Process

After mapping the current process for the product, the team examines the time elements involved in production. Although some of the previous steps can be done in a meeting room, the team must go to the workplace to do time observation.

Key Point

New Tool

The first activity in time observation is measuring the cycle time for each machine operation in the process. The team writes the actions or tasks for one complete machine cycle on the left side of a *Time Observation Sheet* (see Figure 3-5). In addition to actual machine work, a cycle includes other tasks such as loading and unloading, opening and closing machine guards, programming, returning to a neutral position, and other human and machine actions. The team observes the time required for each action during several cycles, then determines an average cycle time for the machine.

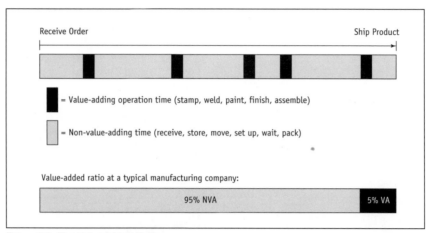

Figure 3-6. The Value-Added Ratio

After observing each operation, the team determines a sample process lead time for the total process. The process lead time includes the cycle time for each operation as well as the time required for transport of WIP and tools between operations.

The team also determines the value-added ratio. The *value-added ratio* is the time spent actually machining or working on the product divided by the total process lead time. Many companies will discover that they have surprisingly low value-added ratios (see Figure 3-6).

TAKE FIVE

Take five minutes to think about these questions and to write down your answers:

- What are some activities in your processing cycle other than the actual value-adding work?

- Do you think your process's value-added ratio is low or high? Why?

Process Capacity Table

	Mgr.	Spvr.	Team	Process Capacity Table	Rev Date: 4/25	Part No: 68745/6	Revision #: ___	Line Name:			Page ___ of ___	
						Part Name: R1/R2		Max. output daily			Current Output/Person/Day	PCS

Step No.	Operation Name	Machine No.	Walk Time	Base Time Manual	Base Time Machine	Base Time Total	Non-Cyclic Tasks Occur.	Non-Cyclic Tasks Time	Non-Cyclic Tasks Time/Pc	Total Time	Daily Capacity	Workers needed (up to 6)	Remarks
1	Rough cut	1		13	(11)	24	0	0	0	24	1150		
2	Cut	2		20	(8)	28	•	•	•	28	985		Remove guards
3	Mill	3		15	(4)	19				19	1452		
4	Drill	4		18	(5)	23				23	1200		Remove guards
5	Gauge	5		14	(10)	24				24	1150		
4A	Outside diameter	4A		18	(5)	23	→	→	→	23	1200		Remove guards
6	Pack	6		18	0	18				18	1533		
													Note: Parts are assembled internal to the machine cycle. Machine time in brackets indicates operator wait time, not machine cycle time.
	Total			116	(43)	159				159			
	Grand Total									159			

Note: All times in seconds. Revise sheet following every improvement and state revision in remarks column.

Figure 3-7. Process Capacity Table

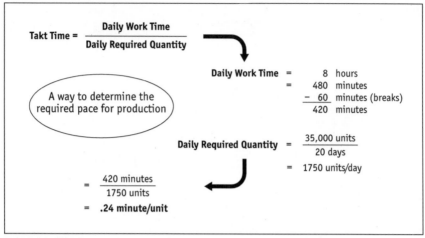

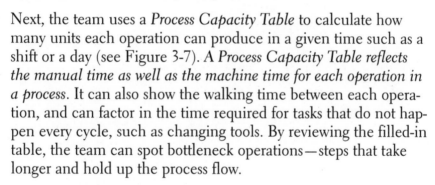

Figure 3-8. Takt Time

Calculate Process Capacity and Takt Time

New Tool

Key Point

Next, the team uses a *Process Capacity Table* to calculate how many units each operation can produce in a given time such as a shift or a day (see Figure 3-7). *A Process Capacity Table reflects the manual time as well as the machine time for each operation in a process.* It can also show the walking time between each operation, and can factor in the time required for tasks that do not happen every cycle, such as changing tools. By reviewing the filled-in table, the team can spot bottleneck operations—steps that take longer and hold up the process flow.

Key Term

Key Point

The team also determines the takt time for the process. *Takt time* is the rate at which each product needs to be completed to meet customer requirements. It is calculated by dividing the daily work time by the daily required quantity (see Figure 3-8). *Takt time, expressed in minutes or seconds per unit, becomes the beat or pulse of the factory. By coordinating process cycle times with takt time, the company can avoid excess product inventory as well as stock outs.*

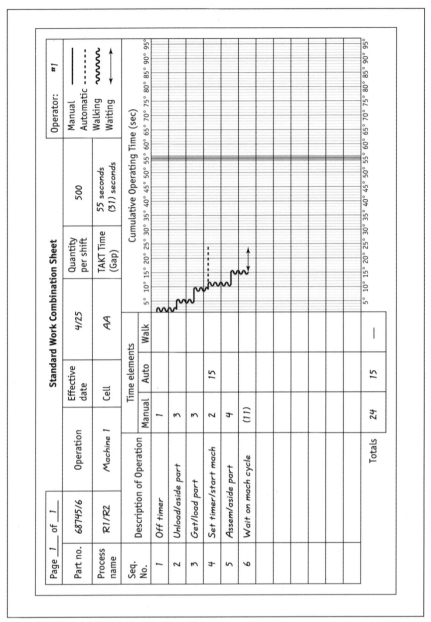

Figure 3-9. Standard Work Combination Sheet

Key Analysis Points

	Before	After
Total square feet	_____	_____
Total distance parts travel	_____	_____
Number of people	_____	_____
Total WIP	_____	_____
Lead time (1 piece)	_____	_____
Value-adding ratio (1 piece)	_____	_____

Figure 3-10. Key Analysis Points

Create Standard Work Combination Sheets

New Tool

Key Point

As the final step in understanding the current conditions, the team prepares a *Standard Work Combination Sheet* for each operation in the process. *This sheet is a graphic display of the relationship between manual work time, machine work time, and walking time for each step in an operation* (see Figure 3-9). The vertical lines represent time. Solid and dotted horizontal lines are drawn to show machine and manual work time; wavy diagonal lines show walking time between steps.

The team also draws a heavy vertical line at the point on the timeline representing the takt time. Operations that extend beyond the takt time may need to be improved so that products can be produced at a rate that meets customer demand.

Key Point

After completing these forms, *the team summarizes key points about the current conditions for later comparison after building the manufacturing cell.* Figure 3-10 shows a sample sheet for displaying this information.

TAKE FIVE

Take five minutes to think about these questions and to write down your answers:

- Which operations have more manual time than machine time?
- Which operations tend to involve a lot of walking or searching for things?

Figure 3-11. The Four Elements of Production

Phase 2: Converting to a Process-Based Layout

Evaluate the Options

Key Point

When it has a grasp on the current situation, the team brainstorms how to rearrange the process elements into a cellular layout to achieve better flow. A useful approach is to *consider how to improve the four basic elements of production: methods, machines, materials, and people* (see Figure 3-11).

Methods

Based on the current methods and sequence shown on the Standard Work Sheet, the team considers new approaches that could promote a better flow and shorten the process lead time.

Machines

The team evaluates machines, equipment, and workstations for movability as well as adaptability to several different products. It also considers ways to use autonomation (*jidoka*), which was described in Chapter 2.

The team also studies other equipment-related issues that may need to be addressed, such as large, unmovable "monument" machines, floor load limitations, supply of utilities such as electricity, air, water, or vacuum, and treatment of waste materials.

Materials

The team examines a standard quantity of materials and WIP that would be needed to operate a process cell. They work for a balanced batch quantity of incoming items to avoid both excess inventory waiting and stock outs.

People

The team considers the number of people available to operate the cell and what training they might need to manage several machines in a cell.

TAKE FIVE

Take five minutes to think about these questions and to write down your answers:

- How do you think the four production elements would need to change to create a manufacturing cell for your process?
- What do you think might be the most important thing to resolve before planning a new layout?

Figure 3-12. Moving the Machines to Create the Cell

Plan Possible New Layouts

When the team has an idea of what to change, it plans a new lay-out. The team follows several guidelines:

Key Points

- Layout in the process sequence is the basic principle.

- Machines are placed close together, with room for only a mini-mum quantity of WIP.

- The layout curves in a U or C shape, with the last machine near the first to reduce walking between cycles.

- The process flow is often counterclockwise. As people walk around to operate the cell, the right hand, which has more control in most people, is then next to the machine; this allows efficient handling of tools and parts, with less turning and reaching over.

Move the Machines

Finally, with a new floor plan in hand, the team is ready to move the machines to create the cell (see Figure 3-12). To make sure this goes smoothly, *the team talks in advance with the people involved in production, maintenance, transport, engineering, and housekeeping, and coordinates the conversion activities with their daily work.*

Key Point

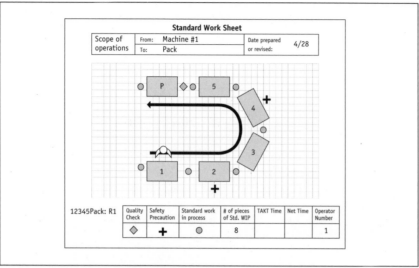

Figure 3-13. Revised Standard Work Sheet

Document the New Operating Procedures

Key Point

After moving the equipment, the team revises the Standard Work Sheet to show the new layout, the amount of WIP in the process, and so on. The new layout is often dramatically different from the old arrangement of machines, people, and WIP (see Figure 3-13).

New Tool

The team also drafts *Standard Operating Procedure* (SOP) sheets that reflect new operating assignments and specific steps in each task. These SOP sheets are used in training the operators to run the new process in the most effective way.

Test to Confirm Improvement

Key Point

Finally, the team tries operating the cell and observes how it works. *The process is timed again to see whether the new layout increases the value-added ratio for the process and meets its takt time.*

TAKE FIVE

Take five minutes to think about these questions and to write down your answers:

• Who should be involved in planning the new process layout? In moving the equipment? In documenting the new procedures and testing them out?

IMPROVEMENT TIPS

1. Focus on improving the standard operations that repeat during the process rather than tasks that happen only occasionally.

2. Use the data on the various standard work forms to identify opportunities for improvement. Look for bottlenecks—operations with longer cycle times, points where WIP gets backed up, stations where defects or equipment failures tend to occur, or long setup times.

3. Be sure to update these sheets after improvement activities to reflect the new situation.

4. Equipment improvement requires time and money. Look first for ways to improve the manual operations in the process.

Phase 3: Continuously Improving the Process

Key Point

Rearranging the shopfloor layout into a manufacturing cell is not really an end point; rather it is the beginning of continuous improvement of the process. Even when the new arrangement dramatically shortens the lead time, most processes can be improved further. This improvement is a continuous effort to fine-tune all aspects of cell operation. Some improvement tips are listed above.

What kind of improvement is appropriate? Look for problems that keep the process from moving in a smooth flow. Common improvement targets include

Example

- Long cycle times
- Product defects
- Long changeover times
- Equipment failures

Improvement approaches for these problems are described in the next sections.

Figure 3-14. Bottlenecks Become Obvious in a Cell

Shorten Cycle Times

Key Point

Once a manufacturing cell is formed, problem operations become very obvious (see Figure 3-14). *When the earlier operations have shorter cycle times, the WIP they produce piles up at the bottleneck process, while the operations after the bottleneck sit idle.*

Example

It's important to look for ways to shorten the processing time, especially for bottleneck operations. An improvement team might review the steps of the operation to see where time could be cut, for example, through better arrangement of tools needed for the operation. It might look at machine settings, such as the time required for the machine to return to the starting position. If the machine cycle time is longer than the takt time, the team may consider adding a second machine of the same kind.

TAKE FIVE

Take five minutes to think about these questions and to write down your answers:

• Which operations are considered bottlenecks in your process? Why do you think this happens?

• What is the usual response to a bottleneck?

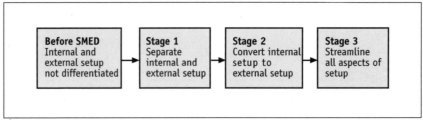

Figure 3-15. The Three Stages of SMED

Shorten Changeover Times

Changeover time is an obstacle to producing several types of products on the same equipment. In the past, companies produced in large lots because setting up for a different product took so much time from production that it wasn't economical unless the cost was spread over a large quantity. However, large-lot production often means large inventories and all the problems, wastes, and costs associated with them.

Key Point

To produce economically in smaller lots, a company must learn how to reduce the time required for changeovers. The ideal is to be able to perform changeover on each machine in the cell within one machine cycle. The important thing is to begin to improve the current times.

Key Term

The single-minute exchange of die* (SMED) *approach* developed by Shigeo Shingo gives a three-stage system for shortening setup (see Figure 3-15).

Stage 1: Separate Internal and External Setup

Key Terms

Internal setup refers to setup operations that can be done only with the equipment stopped. *External setup* operations can be done while the machine is working. At many companies internal and external setup operations are mixed together. This means that things that could be done while the machine is running are not done until the machine is stopped.

---

*Named for the goal of completing changeover within a single-digit number of minutes—9 minutes or fewer.

Stage 1 involves sorting out the external operations so they can be done beforehand. This step alone can reduce setup time by 30 to 50 percent. Typical stage 1 activities include

- Transporting all necessary tools and parts to the machine while it is still running the previous job

- Confirming the function of exchangeable parts before stopping the machine for changeover

Stage 2: Convert Internal Setup to External Setup

The next step is to look again at activities done with the machine stopped and to find ways to do them while the machine is still active. Typical stage 2 improvements include

- Preparing operating conditions in advance, such as preheating a die mold with a heater instead of using trial shots of molten material

- Standardizing functions such as die height to eliminate the need for adjustments

- Using devices that automatically position the parts without measurement

Stage 3: Streamline All Aspects of Setup

This stage chips away at remaining internal setup time in several ways:

- Using parallel operations, with two or more people working simultaneously

- Using functional clamps instead of nuts and bolts

- Using numerical settings to eliminate trial-and-error adjustments

---

TAKE FIVE

Take five minutes to think about these questions and to write down your answers:

- How long does a typical changeover take in your work area?
- Can you list the changeover steps that could take place while the machine is still running?

Zero Quality Control Elements

1. *Source inspection* to catch errors before they become defects

2. *100 percent inspection* to check every workpiece, not just a sample

3. *Immediate feedback* to shorten the time for corrective action

4. *Poka-yoke* (mistake-proofing) devices to check automatically for abnormalities

Figure 3-16. The Four Elements of ZQC

Eliminate Product Defects

Key Point

Product defects hurt the company's reputation with its customers and waste valuable resources in scrap and rework. Companies that pursue low-inventory production no longer have a large buffer to absorb quality defects. To keep production moving smoothly, it is especially important to prevent defects.

Key Point

Key Term

Mistake-proofing is an effective quality assurance approach that prevents defects by catching errors and other nonstandard conditions before they actually turn into defects. The mistake-proofing system known as *Zero Quality Control* (ZQC) ensures zero defects by inspecting the processing conditions for 100 percent of the work, ideally just before an operation is performed. If an error is discovered, the process shuts down and gives immediate feedback with lights, warning sounds, and so on. The basic elements of the ZQC approach to mistake-proofing are summarized in Figure 3-16.

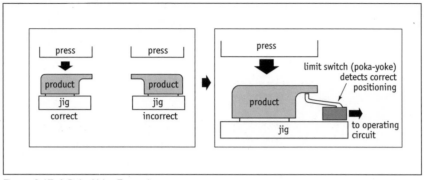

Figure 3-17. A Poka-Yoke Example

Key Term

Because people can make mistakes even in inspection, mistake-proofing often relies on sensing mechanisms called *poka-yoke*, which check conditions automatically and signal when problems occur. Poka-yoke devices include electronic sensors such as limit switches and photoelectric eyes, as well as passive devices such as positioning pins that prevent backward insertion of a workpiece. Figure 3-17 shows an example of a limit switch used as a poka-yoke device to prevent processing when the work is placed incorrectly. Poka-yoke devices may use counters to make sure an operation is repeated the correct number of times.

Key Point

The key to effective mistake-proofing is determining when and where defect-causing conditions arise and then figuring out how to detect or prevent these conditions, every time. Shopfloor people have important knowledge and ideas to share for developing and implementing poka-yoke systems that check every item and give immediate feedback about the problem.

TAKE FIVE

Take five minutes to think about these questions and to write down your answers:

- What kinds of actions or conditions can cause defects in your process? Can you think of ways to catch these conditions before defects are made?

Figure 3-18. A Definition of TPM

Reduce Equipment Failures

Key Point

The health of processing and assembly equipment can make or break a cellular manufacturing effort. *Since it does not have large WIP buffers between processes, cellular manufacturing requires dependable equipment that will perform as needed, when needed.*

Key Term

Total productive maintenance (TPM) is a good way to ensure that equipment is ready when it is needed. TPM is a comprehensive, companywide approach for reducing equipment-related losses such as downtime, speed reduction, and defects by stabilizing and improving equipment conditions. The definition in Figure 3-18 describes five key aspects.

Key Point

The TPM framework improves equipment effectiveness through various approaches that involve everyone in the company. Frontline workers, especially, have a key role in the TPM activity called autonomous maintenance. This ongoing activity is described further in Chapter 4.

TAKE FIVE

Take five minutes to think about these questions and to write down your answers:

- Are there machines in your work area that tend to fail or have minor stoppages? What is the usual response when one of these events happens?

In Conclusion

SUMMARY

The conversion from an operation-based factory layout to cell manufacturing has three basic phases:

- Understanding the current conditions
- Converting to a process-based layout
- Continuously improving the process

Phase 1, understanding the current conditions on the shop floor, gives the team involved in the cell conversion a baseline for measuring improvement. The team collects data such as the product mix (using PQ analysis) and resources available for production. The team also documents the layout and flow of the current process, using process route analysis to determine product families and drawing a process map on a Standard Work Sheet.

Next the team times the process, recording the time for the steps in one cycle of each machine in the process. The team uses the cycle times for the operations in the process to determine a sample lead time and value-added ratio for the total process. It then uses a Process Capacity Table to calculate how many units each individual operation can produce in a given time.

For the Process Capacity Table the team determines the takt time for the process. Takt time is the rate at which each product needs to be completed to meet customer requirements. It is calculated by dividing the daily work time by the daily required quantity. Takt time, expressed in seconds per unit, becomes the beat or pulse of the factory.

The final step in understanding the current conditions is creation of a Standard Work Combination Sheet for each operation in the process. This sheet is a graphic display of the time spent in manual work, machine work, and walking for each step in a particular operation. The team indicates the takt time on this graph to show when the actual cycle time needs to be shortened to match demand.

Phase 2, conversion to a process-based layout, begins with an evaluation of ways to improve the four basic elements of production: methods, machines, materials, and people. Based on the current methods and sequence shown on the Standard Work Sheet, the team considers new methods to promote a better flow and shorten process lead time.

The team evaluates machines, equipment, and workstations for movability and adaptability. It considers ways to use autonomation (*jidoka*) and studies other equipment-related issues that may need to be addressed, such as unmovable "monument" machines, floor load limitations, utilities, and waste treatment.

The team determines the quantity of materials and WIP required to operate the cell. It considers the number of people available to operate the cell and the training they might need to manage several machines.

When the team has an idea of what it would like to see happen, it plans a new layout to serve those needs. The plan places machines in the process sequence, close together for minimum WIP, and usually in a U or C shape so the operator need not walk far to begin the next cycle. Then the team discusses the plan with the other functions involved and coordinates with them to move the equipment.

With the equipment in place in the new cell, the team revises the Standard Work Sheets to show the new layout, the amount of WIP in the process, and so on. Then the team drafts Standard Operating Procedure sheets to reflect assignments and steps in the revised process. Finally, the team tries operating the cell and observes how it works.

Phase 3, continuously improving the process, begins when the new layout is in place. Rearranging the shopfloor layout into a manufacturing cell is the beginning of continuous improvement of the process. Even when the new arrangement dramatically shortens lead time, most processes can always be improved further. Common improvement targets include

- long cycle times
- product defects
- long changeover times
- equipment failures

In cellular manufacturing, it's important to shorten processing time, especially for bottleneck operations. An improvement team might look at factors such as how tools are arranged or how long it takes the machine to return to the starting position.

Changeover time is an obstacle to producing several types of products on the same equipment. The single-minute exchange of die (SMED) approach gives a three-stage system for shortening setup.

Defect prevention is especially important for smooth production. Mistake-proofing is an effective approach that prevents defects by catching errors and nonstandard conditions before they turn into defects. Because people can make mistakes even in inspection, mistake-proofing often relies on sensing mechanisms called *poka-yoke*, which check conditions automatically and signal when problems occur. The key to effective mistake-proofing is determining when and where defect-causing conditions arise and then figuring out how to detect or prevent these conditions, every time.

Cellular manufacturing requires dependable equipment. Total productive maintenance (TPM) is a comprehensive, company-wide approach for reducing equipment-related losses such as downtime, speed reduction, and defects by stabilizing and improving equipment conditions. TPM improves equipment effectiveness through approaches that involve everyone in the company.

REFLECTIONS

Now that you have completed this chapter, take five minutes to think about these questions and to write down your answers:

- What did you learn from reading this chapter that stands out as particularly useful or interesting?

- Do you have any questions about the topics presented in this chapter? If so, what are they?

- What additional information do you need to fully understand the ideas presented in this chapter?

Chapter 4

Teamwork Tools for Cellular Manufacturing

Working in Teams

The success of cellular manufacturing depends on teamwork. In a cell, people often work together in new ways. For example, in large-lot manufacturing, an operator might run just one type of machine, with the goal of producing a specified output quantity.

Key Point

In a cell, however, several different operations are combined in a sequence, and the main job of the people working in the cell is to maintain a smooth flow through all the operations. This requires people to coordinate their work in the cell. Employees working in cells often learn several operations so they can proactively keep the work flowing.

Key Point

Teamwork is also important for improvement activities. *A group of employees has more creative potential and energy than any person working on a problem alone.* The next several pages introduce some team-based improvement approaches you may use in a cellular manufacturing situation.

Standardizing Workplace Conditions through 5S

Cellular manufacturing cannot succeed in a workplace that is cluttered, disorganized, or dirty. Poor workplace conditions lead to wastes such as extra motion to avoid obstacles, time spent searching for things, and delays due to defects, machine failures, or accidents.

Key Point

Establishing basic workplace conditions is an essential first step in creating a manufacturing cell. In many companies, employee teams use the 5S system to improve and standardize workplace conditions for safe and effective operation.

Key Term

The 5S system is a set of five basic principles that have names beginning with S:

- Sort
- Set in Order
- Shine
- Standardize
- Sustain

Key Terms

Sort: The team begins by sorting out and removing items not needed in the work area. Teams often use the Red-Tag technique to identify unneeded items and manage their disposition.

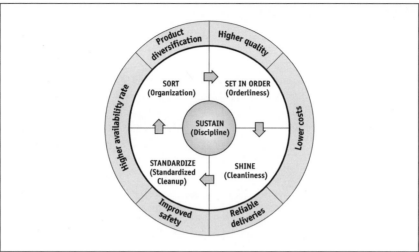

Figure 4-1. The 5S System

Key Terms

Set in Order: Next, the team members determine appropriate locations for needed items, then apply temporary lines, labels, and signboards to indicate the new locations. The main idea is, "A place for everything, and everything in its place."

Shine: The third S involves a top to bottom cleaning of the work area, including the equipment. Shine also means inspecting equipment during cleaning to spot early signs of trouble.

Standardize: In the fourth S, the team establishes the improved conditions as workplace standards. Visual management methods are adopted to ensure that everyone understands and can easily follow the new standards.

Sustain: The final 5S principle uses training and communication to maintain and monitor the improved conditions and to spread 5S ideas and activities to other areas of the company.

TAKE FIVE

Take five minutes to think about these questions and to write down your answers:

- What conditions would you change to make your work area easier to use?
- What advantages do you see from doing 5S activities as a team rather than individually?

Figure 4-2. An Andon Board

Using Visual Management for Production Control and Safety

Key Point

Visual management is an important support for cellular manufacturing. *Visual management techniques express information in a way that can be understood quickly by everyone.* Sharing information through visual tools helps keep production running smoothly and safely. Shopfloor teams are often involved in devising and implementing these tools through 5S and other improvement activities.

Key Term

One form of visual management often seen in manufacturing cells is the *andon system*. In an andon system, individual machines or assembly stations are equipped with call lamps. If the machine breaks down or runs out of parts, the operator (or the machine itself) turns on the light to call attention. At many plants, overhead andon boards also show the status of several machines or lines to help others locate the problem (see Figure 4-2).

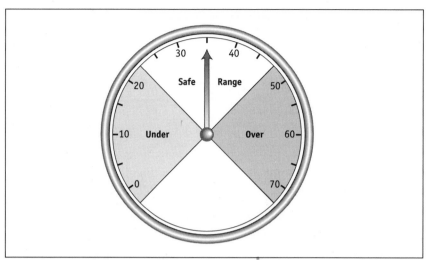

Figure 4-3. A Color Coded Dial Gauge

Key Point

Visual location indicators help keep order in the workplace.
Lines, labels, and signboards (used during 5S implementation)
tell everyone at a glance where to find things and where to put
them away. Using these methods can eliminate a lot of time
wasted in searching.

Key Point

X

Example

Visual information can also help prevent mistakes. Color coding is
a form of visual display often used to prevent errors. Shaded "pie
slices" on a dial gauge tell the viewer instantly when the needle is
out of the safe range (see Figure 4-3). Matching color marks is
another approach that can help people use the right tool or
assemble the right part.

TAKE FIVE

Take five minutes to think about these questions and to write
down your answers:

- Can you find example of visual displays already used in your
  workplace?
- What are some other visual methods you could use to reduce
  waste and errors?

Figure 4-4. Autonomous Maintenance Involves Everyone

Performing Autonomous Maintenance Activities

Key Term

Autonomous maintenance refers to activities carried out by shopfloor teams in cooperation with maintenance staff to help stabilize basic equipment conditions and spot problems early. Autonomous maintenance—an element of total productive maintenance (TPM)—changes the old view that operators just run machines and maintenance people just fix them. Operators have valuable knowledge and skill that can be used to help keep equipment from breaking down.

Key Point

In autonomous maintenance, operators learn how to clean the equipment they use every day, and how to inspect for trouble signs as they clean (see Figure 4-4). They may also learn basic lubrication routines, or at least how to check for adequate lubrication. They learn simple methods to reduce contamination and keep the equipment cleaner. Ultimately, they learn more about the various operating systems of the equipment and may assist technicians with repairs.

Step 1. Conduct initial cleaning and inspection.

Step 2. Eliminate sources of contamination and inaccessible areas.

Step 3. Develop and test provisional cleaning, inspection, and lubrication standards.

Step 4. Conduct general inspection training and develop inspection procedures.

Step 5. Conduct general inspections autonomously.

Step 6. Apply standardization and visual management throughout the workplace.

Step 7. Conduct ongoing autonomous maintenance and advanced improvement activities.

Figure 4-5. Autonomous Maintenance Activities

Autonomous maintenance is, at its heart, a team-based activity. Through the steps of autonomous maintenance activities, shopfloor employees work with maintenance technicians and engineers toward a common goal—more effective equipment (see Figure 4-5). By sharing what they know, they can catch many of the problems that cause failures, defects, or accidents.

TAKE FIVE

Take five minutes to think about these questions and to write down your answers:

- Who performs basic cleaning and maintenance on the equipment in your work area?
- Do you think autonomous maintenance activities would reduce unplanned downtime in your company? Why or why not?

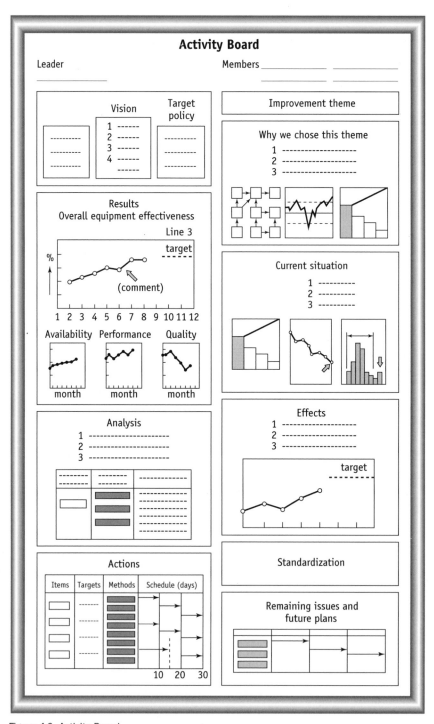

Figure 4-6. Activity Board

Using Activity Boards and One-Point Lessons

Sharing information is an important element of teamwork. Activity boards and one-point lessons are two useful approaches for making information public.

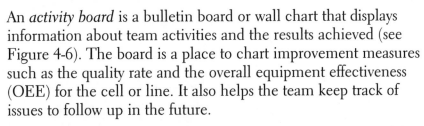

Key Term

An *activity board* is a bulletin board or wall chart that displays information about team activities and the results achieved (see Figure 4-6). The board is a place to chart improvement measures such as the quality rate and the overall equipment effectiveness (OEE) for the cell or line. It also helps the team keep track of issues to follow up in the future.

Key Term

A *one-point lesson* is an easy-to-read poster or sheet developed by team members to teach others about a particular problem, an improvement example, or basic knowledge that everyone should have. The lesson is kept short and focused on one point so people can get the information quickly. One-point lessons are often illustrated with photographs or drawings of the situation. Figure 4-7 on page 58 shows an example of a one-point lesson describing a problem and actions taken to prevent it.

TAKE FIVE

Take five minutes to think about these questions and to write down your answers:

- What kinds of information would be useful to display on an activity board in your work area?
- What topic would you choose for a one-point lesson for your work area?

	One-Point Lesson
	Changeover on the P3 Machine
1 Prepare!	Gather all implements and the new machine tool in advance, during production of the current product.
2	Put the new machine tool into a spare holder cartridge until it clicks. Be sure the line on the side of the tool aligns with the line on the cartridge.
3 (OFF)	When the current product run ends, take the machine offline and remove the old cartridge.
4 Click!	Wipe off the tool stem on the machine, using a clean rag, then click the new cartridge into place.
5 (ON)	Restart the machine for the next production run.
6 Check!	Remove the old tool from its cartridge. Put the tool in the recycle bin. Check the cartridge for damage. If the cartridge is undamaged, store it for the next changeover.

Figure 4-7. One-Point Lesson

In Conclusion

SUMMARY

Successful cellular manufacturing depends on teamwork. The main job of the people working in a cell is to maintain a smooth flow through all the operations. This requires people to coordinate their work. Employees often learn several operations so they can keep the work flowing. Teamwork is also important for improvement activities. A group of employees has more creative potential and energy than any person working on a problem alone.

Establishing basic workplace conditions is an essential first step in creating a manufacturing cell. In many companies, employee teams use the 5S system to improve and standardize workplace conditions for safe and effective operation. The 5S system is a set of five basic principles:

Sort: Sorting out and removing items not needed in the work area, often using the Red-Tag technique.

Set in Order: Determining appropriate locations for needed items, then applying temporary lines, labels, and signboards to indicate the new locations. The concept: "A place for everything, and everything in its place."

Shine: Thoroughly cleaning the work area and inspecting equipment to spot early signs of trouble.

Standardize: Establishing the improved conditions as workplace standards and applying visual management methods.

Sustain: Using training and communication to maintain and monitor the improved conditions and to spread 5S ideas and activities.

Visual management techniques express information in a way that can be understood quickly by everyone. Sharing information through visual tools helps keep production running smoothly and safely. Shopfloor teams are often involved in devising and implementing these tools through 5S and other improvement activities. Andon lights and boards are one form of visual management that displays machine status for easy location of problems. Visual location indicators such as lines, labels, and signboards can eliminate a lot of time wasted in searching. Color coding and other visual information can also help prevent mistakes.

Autonomous maintenance refers to shopfloor team activities to help stabilize basic equipment conditions and spot problems early. Autonomous maintenance—an element of total productive maintenance (TPM)—changes the old view that operators just run machines and maintenance people just fix them. In autonomous maintenance, operators learn how to clean the equipment and how to inspect for trouble signs as they clean. They may also learn basic lubrication routines and simple methods to reduce contamination and keep equipment cleaner.

Sharing information is an important element of teamwork. Activity boards and one-point lessons are two useful approaches for making information public.

An activity board is a bulletin board or wall chart that displays information about team activities and the results achieved. The board is a place to chart improvement measures such as the quality rate and the overall equipment effectiveness (OEE) for the cell or line. It also helps the team track issues to follow up.

A one-point lesson is an easy-to-read poster or sheet developed by team members to teach others about a particular problem, an improvement example, or basic knowledge that everyone should have. The lesson is kept short and focused on one point so people can get the information quickly; photographs or drawings are often used.

REFLECTIONS

Now that you have completed this chapter, take five minutes to think about these questions and to write down your answers:

• What did you learn from reading this chapter that stands out as particularly useful or interesting?

• Do you have any questions about the topics presented in this chapter? If so, what are they?

• What additional information do you need to fully understand the ideas presented in this chapter?

Chapter 5

Reflections and Conclusions

CHAPTER OVERVIEW

Reflecting on What You've Learned

Opportunities for Further Learning

Conclusions

Additional Resources Related to Cellular Manufacturing

> Books and Videos
>
> Newsletters
>
> Training and Consulting
>
> Website

Reflecting on What You've Learned

Key Point

An important part of learning is reflecting on what you've learned. Without this step, learning can't take place effectively. That's why we've asked you at the end of each chapter to reflect on what you've learned. And now that you've reached the end of the book, we'd like to ask you to reflect on what you've learned from the book as a whole.

- Take ten minutes to think about the following questions and to write down your answers.

- What did you learn from reading this book that stands out a particularly useful or interesting?

- What ideas, concepts, and techniques have you learned that will be most useful to you during implementation of cellular manufacturing? How will they be useful?

- What ideas, concepts, and techniques have you learned that will be least useful during implementation of cellular manufacturing? Why won't they be useful?

- Do you have any questions about cellular manufacturing? If so, what are they?

Opportunities for Further Learning

Here are some ways to learn more about cellular manufacturing:

How-to Steps

- Find other books, videos, or trainings on this subject. Several are listed on the next pages.

- If your company is already implementing cellular manufacturing, visit other departments or areas to see how they are applying the ideas and approaches you have learned about here.

- Find out how other companies have implemented cellular manufacturing. You can do this by reading magazines and books about manufacturing cells, lean manufacturing, or just-in-time manufacturing, and by attending conferences and seminars presented by others.

Conclusions

Cellular manufacturing is more than a series of techniques. It is a fundamental approach to improving the manufacturing process. We hope this book has given you a taste of how and why this approach can be helpful and effective for you in your work.

Additional Resources Related to Cellular Manufacturing

Books and Videos

One-Piece Flow, Just-in-Time, and Lean Manufacturing

Ken'ichi Sekine, *One-Piece Flow: Cell Design for Transforming the Manufacturing Process* (Productivity Press, 1992)—This comprehensive book describes how to redesign the factory layout for most effective deployment of equipment and people; it includes many examples and illustrations.

Jeffrey Liker, *Becoming Lean: Inside Stories of U.S. Manufacturers* (Productivity Press, 1997)—This book shares powerful first-hand accounts of the complete process of implementing cellular manufacturing, just-in-time, and other aspects of lean production.

Japan Management Association, ed., *Kanban and Just-in-Time at Toyota* (Productivity Press, 1986)—This classic overview book describes the underlying concepts and main techniques of the original lean manufacturing system.

Taiichi Ohno, *Toyota Production System* (Productivity Press, 1988)—This is the story of the first lean manufacturing system, told by the Toyota vice president who was responsible for implementing it.

Hiroyuki Hirano, *JIT Factory Revolution* (Productivity Press, 1988)—This book of photographs and diagrams gives an excellent overview of the changes involved in implementing a lean, cellular manufacturing system.

Hiroyuki Hirano, *JIT Implementation Manual* (Productivity Press, 1990)—This two-volume manual is a comprehensive, illustrated guide to every aspect of the lean manufacturing transformation.

Shigeo Shingo, *A Study of the Toyota Production System from an Industrial Engineering Viewpoint* (Productivity Press, 1989)—This classic book was written by the renowned industrial engineer who helped develop key elements of its success.

Iwao Kobayashi, *20 Keys to Workplace Improvement* (Productivity Press, 1995)—This book addresses 20 key areas in which a company must improve to maintain a world class manufacturing operation. A five-step improvement for each key is described and illustrated.

The 5S System and Visual Management

The 5S System (Tel-A-Train, 1997)—Filmed at leading U.S. companies, this seven-tape training package (coproduced with Productivity Press) teaches shopfloor teams how to implement the 5S System.

Productivity Press Development Team, *5S for Operators* (Productivity Press, 1996)—This Shopfloor Series book outlines five key principles for creating a clean, visually organized workplace that is easy and safe to work in. Contains numerous tools, illustrated examples, and how-to steps, as well as discussion questions and other learning features.

Michel Greif, *The Visual Factory: Building Participation Through Shared Information* (Productivity Press, 1991)—This book shows how visual management techniques can provide "just-in-time" information to support teamwork and employee participation, on the factory floor.

Poka-Yoke (Mistake-Proofing) and Zero Quality Control

Productivity Press Development Team, *Mistake-Proofing for Operators* (Productivity Press, 1997)—This Shopfloor Series book describes the basic theory behind mistake-proofing and introduces *poka-yoke* systems for preventing errors that lead to defects.

Shigeo Shingo, *Zero Quality Control: Source Inspection and the Poka-Yoke System* (Productivity Press, 1986)—This classic book tells how Shingo developed his ZQC approach. It includes a detailed introduction to *poka-yoke* devices and many examples of their application in different situations.

NKS/Factory Magazine, ed., *Poka-Yoke: Improving Product Quality by Preventing Defects* (Productivity Press, 1988)—This illustrated book shares 240 poka-yoke examples implemented at different companies to catch errors and prevent defects.

Quick Changeover

Productivity Press Development Team, *Quick Changeover for Operators* (Productivity Press, 1996)—This Shopfloor Series book describes the stages of changeover improvement with examples and illustrations.

Shigeo Shingo, *A Revolution in Manufacturing: The SMED System* (Productivity Press, 1985)—This classic book tells the story of Shingo's SMED System, describes how to implement it, and provides many changeover improvement examples.

Total Productive Maintenance

Japan Institute of Plant Maintenance, ed., *TPM for Every Operator* (Productivity Press, 1996)—This Shopfloor Series book introduces basic concepts of TPM, with emphasis on the six big equipment-related losses, autonomous maintenance activities, and safety.

Japan Institute of Plant Maintenance, ed., *Autonomous Maintenance for Operators* (Productivity Press, 1997)—This Shopfloor Series book on key autonomous maintenance activities includes chapters on cleaning/inspection, lubrication, localized containment of contamination, and one-point lessons related to maintenance.

Newsletters

Lean Manufacturing Report—News and case studies on how companies are implementing lean manufacturing philosophy and specific techniques such as cell design and total productive maintenance. For subscription information, call 1-800-394-6868.

Training and Consulting

Productivity Consulting Group offers a full range of consulting and training services on lean manufacturing approaches and cell design. For additional information, call 1-800-394-6868.

Website

Visit our web pages at *www.productivityinc.com* to learn more about Productivity's products and services related to cellular manufacturing.

About the Productivity Development Team

Since 1979, Productivity, Inc. has been publishing and teaching the world's best methods for achieving manufacturing excellence. At the core of this effort is a team of dedicated product developers, including writers, instructional designers, editors, and producers, as well as content experts with years of experience in the field. Hands-on experience and networking keep the team in touch with changes in manufacturing as well as in knowledge sharing and delivery. The team also learns from customers and applies this knowledge to create effective vehicles that serve the learning needs of every level in the organization.

LEARNING PACKAGE

The Learning Package is designed to give your team leaders everything they need to facilitate study groups on *Cellular Manufacturing*. Shopfloor workers participate through a series of discussion and application sessions that highlight the tools and techniques they've learned from the book.

The Learning Package:

- Provides the foundation for launching a full-scale implementation process
- Provides immediate practical skills for participants
- Offers a flexible course design you can adapt to your unique requirements
- Encourages workers to become actively involved in their own learning process

Included in the Learning Package:

- Five copies of *Cellular Manufacturing*
- One copy of *One-Piece Flow*
- One 8-1/2" x 11" Leader's Guide
- A set of overhead transparencies that summarize major points
- A set of slides with case study examples

Cellular Manufacturing Learning Package
The Productivity Development Team
ISBN 1-56327-214-8
Order CELLLP-B8009 / $295.00

About the Shopfloor Series

Put powerful and proven improvement tools in the hands of your entire workforce!

Progressive shopfloor improvement techniques are imperative for manufacturers who want to stay competitive and to achieve world class excellence. And it's the comprehensive education of all shopfloor workers that ensures full participation and success when implementing new programs. The Shopfloor Series books make practical information accessible to everyone by presenting major concepts and tools in simple, clear language and at a reading level that has been adjusted for operators by skilled instructional designers. One main idea is presented every two to four pages so that the book can be picked up and put down easily. Each chapter begins with an overview and ends with a summary section. Helpful illustrations are used throughout.

Books currently in the Shopfloor Series include:

5S FOR OPERATORS
5 Pillars of the Visual Workplace
The Productivity Press Development Team
ISBN 1-56327-123-0 / 133 pages
Order 5SOP-B8009 / $25.00

QUICK CHANGEOVER FOR OPERATORS
The SMED System
The Productivity Press Development Team
ISBN 1-56327-125-7 / 93 pages
Order QCOOP-B8009 / $25.00

MISTAKE-PROOFING FOR OPERATORS
The Productivity Press Development Team
ISBN 1-56327-127-3 / 93 pages
Order ZQCOP-B8009 / $25.00

JUST-IN-TIME FOR OPERATORS
The Productivity Press Development Team
ISBN 1-56327-134-6 / 96 pages
Order JITOP-B8009 / $25.00

TPM FOR EVERY OPERATOR
The Japan Institute of Plant Maintenance
ISBN 1-56327-080-3 / 136 pages
Order TPMEO-B8009 / $25.00

TPM FOR SUPERVISORS
The Productivity Press Development Team
ISBN 1-56327-161-3 / 96 pages
Order TPMSUP-B8009 / $25.00

TPM TEAM GUIDE
Kunio Shirose
ISBN 1-56327-079-X / 175 pages
Order TGUIDE-B8009 / $25.00

AUTONOMOUS MAINTENANCE
The Japan Institute of Plant Maintenance
ISBN 1-56327-082-x / 138 pages
Order AUTOMOP-B8009 / $25.00

FOCUSED EQUIPMENT IMPROVEMENT FOR TPM TEAMS
The Japan Institute of Plant Maintenance
ISBN 1-56327-081-1 / 144 pages
Order FEIOP-B8009 / $25.00

Productivity Press, Dept. BK, P.O. Box 13390, Portland, OR 97213-0390
Telephone: **1-800-394-6868** Fax **1-800-394-6286**

CONTINUE YOUR LEARNING WITH IN-HOUSE TRAINING AND CONSULTING FROM PRODUCTIVITY, INC.

Productivity, Inc. offers a diverse menu of consulting services and training products based on the exciting ideas contained in the books of Productivity, Inc. Whether you need assistance with long-term planning or focused, results-driven training, Productivity's experienced professional staff can enhance your pursuit of competitive advantage.

Productivity, Inc. integrates a cutting edge management system with today's leading process improvement tools for rapid, measurable, lasting results. In concert with your management team, we will focus on implementing the principles of Value-Adding Management, Total Quality Management, Just-in-Time, and Total Productive Maintenance. Each approach is supported by Productivity's wide array of team-based tools: Standardization, One-Piece Flow, Hoshin Planning, Quick Changeover, Mistake-Proofing, Kanban, Problem Solving with CEDAC, Visual Workplace, Visual Office, Autonomous Maintenance, Equipment Effectiveness, Design of Experiments, Quality Function Deployment, Ergonomics, and more.

Productivity is known for significant improvement on the shop floor and the bottom line. Through years of repeat business, an expanding and loyal client base continues to recommend Productivity to their colleagues. Contact us to learn how we can tailor our services to fit your needs.

Telephone: 1-800-394-6868 (U.S. only) or 1-503-235-0600
Fax: 1-800-394-6286

THE SHOPFLOOR SERIES LEARNING ASSESSMENT PACKAGE

Software to Confirm the Learning of Your Knowledge Workers

Created by the Productivity Development Team

How do you know your employee education program is getting results? Employers need to be able to quantify the benefit of their investment in workplace education. The *Shopfloor Series books* and *Learning Packages* from Productivity Press offer a simple, cost-effective approach for building basic knowledge about key manufacturing improvement topics. Now you can confirm the learning with the *Shopfloor Series Learning Assessment*.

The *Shopfloor Series Learning Assessment* is a new software package developed specifically to complement five key books in the *Shopfloor Series*. Each module of the Learning Assessment provides knowledge tests based on the contents of one of the *Shopfloor Series books*, which are written for production workers. After an employee answers the questions for a chapter in the book, the software records his or her score. Certificates are included for recognizing the employee's completion of the assessment for individual modules and for all five core modules.

The *Shopfloor Series Learning Assessment* will help your company ensure that employees are learning and are recognized and rewarded for gaining knowledge. It supports professional development for your employees as well as effective implementation of shopfloor improvement programs.

ISBN 1-56327-203-2
Order ASSESS-B8009 / $1495.00

Here's How the Learning Assessment Package Works:

1. The employee reads one of the Shopfloor Series books, chapter by chapter. Easy to read and understand, the books educate your employees with information they need, and prepare them for the learning assessment test questions.

2. After an administrator has set up the Learning Assessment software on a computer, the employee can then use the computer to answer a set of test questions about the information in the Shopfloor Series book they have read. The software automatically scores the answers and logs the score into a database for easy access by the administrator.

3. If the employee does not pass the assessment for a particular chapter, he or she can review the material in the book and take the assessment again. (For security, the software selects randomly from three different questions on each topic.)

4. Upon passing the assessment modules for all chapters of the Shopfloor Series books, the employee receives a completion certificate (included in the package) and any other reward or recognition determined by your company.